THIS BOOK BELONGS TO

_____

MAKER

Make:
WOODWORKING FOR YOUNG MAKERS

FUN AND EASY DO-IT-YOURSELF PROJECTS

LOYD BLANKENSHIP AND LANE BOYD ★ ART BY MIKE GRAY

MAKER MEDIA
San Francisco, CA

Woodworking for Young Makers
Fun and Easy Do-It-Yourself Projects

Loyd Blankenship and Lane Boyd

Art by Mike Gray

Published by Maker Media, Inc., 1160 Battery Street East, Suite 125, San Francisco, California 94111.

Maker Media books may be purchased for educational, business, or sales promotional use. Online editions are also available for most titles (safaribooksonline.com). For more information, contact our corporate/institutional sales department: 800-998-9938 or corporate@oreilly.com.

Publisher and Editor: Roger Stewart
Art Director, Designer, and Compositor: Jason Babler
Cover Designers: Jason Babler, Maureen Forys

Previously published as *Making a Wizard's Wand,
a Secret Compartment Box, and More!*,
November 2016.

Revision History for This Edition

2017-04-15 First Release

978-1-680-45281-5

To the memory of my father-in-law,
John Samuel Gordon Jr., 1940-2015.
He was a talented and passionate
woodworker and is sorely missed.

–Loyd

For my mom, who made the magic possible.

–Lane

For my Olivia, you are the best daughter in the
world, and I love you more than anything.

–Mike

Loyd Blankenship has more than two decades of woodworking experience and, when not busy with his day job at Intel, can often be found making stuff at Ocean State Maker Mill in Rhode Island. He has written on topics ranging from sports to computers to roleplaying games.

———□———

Lane Boyd is a former political beat reporter and computer graphic design journalist who now manages websites and email campaigns for the healthcare industry. You can find more of his wand-making tutorials at spellbooknotincluded.com.

———□———

Mike Gray is an award-winning writer, cartoonist, animator, and director. He is also a proud father but, so far, has not won an award for that. You can peruse his portfolio online at pencilforhire.com.

Table of Contents

Welcome to the World of Woodworking! 1

    Be Safe! 2

Make a Wizard's Wand 4

    All About Sandpaper 16

    Essential Tools 19

Make a Sanding Block 22

    All About Glue 31

Make a Charging Stand 34

    Finishing Your Wood 48

    Joinery: The Art of Holding Wood Together 50

Make a Secret-Compartment Box 56

    Common Woodworking Woods 75

What's Next? 77

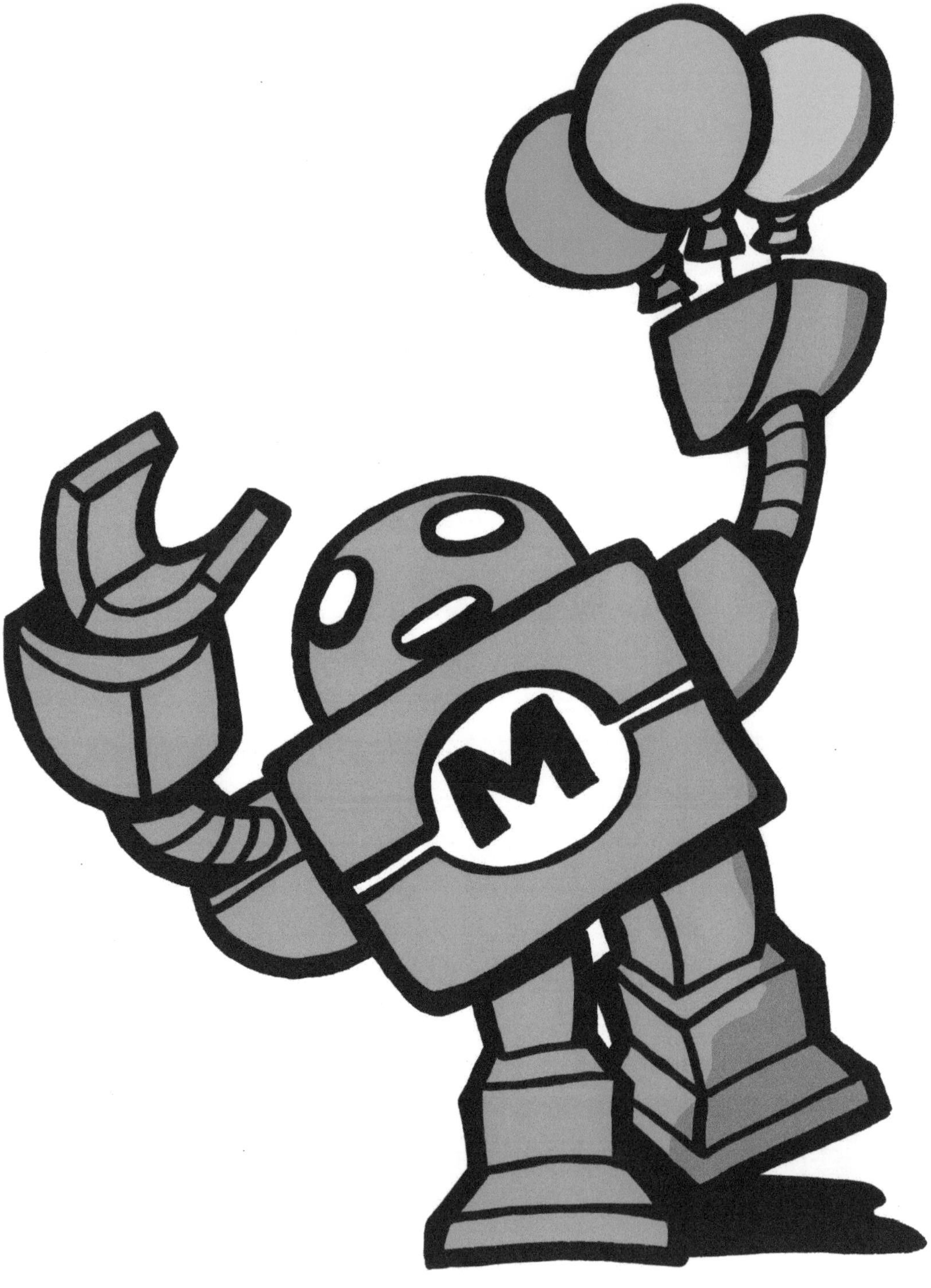

WELCOME TO THE WORLD OF WOODWORKING!

Hi! I'm Makey!

You may have seen me if you've been to a Maker Faire. I was created to make, and I love to help people become Makers.

In this book, we're going to work together on several woodworking projects that are fun and will have your friends and family saying, "How did you make that?!"

You'll learn how to use common workshop tools to make uncommon stuff. I'll show you how to use tools safely and correctly and how choosing the right materials can add that wow factor to your finished pieces. I'll even share some pro Maker tips along the way!

You'll find that as you work on these projects, your woodworking skills and imagination will grow and you'll probably start to think of new things to make on your own. Making with wood can be a hobby—or maybe even a profession—that will stick with you your whole life.

Let's get making!

BE SAFE!

We're mostly going to use hand tools — such as hammers, carving knives, and handsaws — but we'll also use a couple of power tools. You can hurt yourself with both kinds of tools if you handle them carelessly, so be safe and no goofing around!

THE EYES HAVE IT

You will definitely want to finish a project with the same number of working eyes you had when you started! The first thing you should do when you step into the shop, then, is to put on safety glasses.

DON'T STAIN YOUR HANDS

Whenever you apply a finish or stain to wood, wear a pair of cheap, disposable nitrile gloves. You can find these at most hardware stores.

WHAT *NOT* TO WEAR

Wearing anything "dangly" is a bad idea around drills and rotary tools. This applies not just to loose clothing, bracelets, and necklaces, but to long hair, too. Consider pulling it back or tucking it under a hat when you're in the shop. If you are wearing a long-sleeved shirt, make sure the cuffs are tight.

DUST IN THE WIND

Whenever you use sandpaper, you should wear a dust mask. You can find inexpensive disposable ones at most hardware stores.

DON'T POINT THE SHARP BITS AT YOUR BODY

When using sharp objects, don't angle the cutting edges or pointy bits toward your hands or body. Wear a glove to protect your hand when carving with a knife or rotary tool.

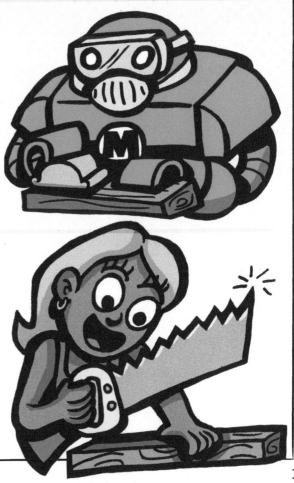

MAKE A
WIZARD'S WAND

Do you like stories about fantastic adventures, fabulous creatures, and magic artifacts? ⚡ Let's make a wizard's wand that you can use with costume play, as a prop for a play, or just as a fun decoration for your room. ⚡ Making a wizard's wand is a great place to begin woodworking because you can make it as simple or as fancy as you want. ⚡

DO IT YOUR WAY!

There is no right or wrong way when it comes to crafting a wand. In later projects you will have to be precise in your cuts and measurements. Here, though, the only limits are your imagination!

GETTING STARTED

For this project, you can choose to use a wood carving (or whittling) knife or a rotary tool. Either will work, and both are fun to use!

WOOD CARVING KNIVES

You can find many at arts & crafts stores and hardware stores. Use a steel blade that is less than 4 inches long and fairly narrow (3/4 inch or smaller) for fitting into tight spaces.

TOOLBOX

WOOD CARVING KNIFE

OR

ROTARY TOOL

HANDSAW

OR

DOVETAIL SAW

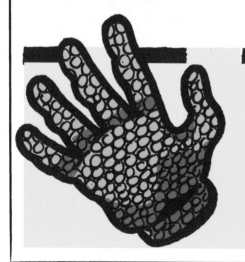

STAY SAFE!

When working with any sharp tool, you should be extra careful. Cut away from your hands and body. Wear a protective glove on the hand that is holding the wand while you carve. And have an adult around at all times!

MAKE A WIZARD'S WAN

MATERIALS

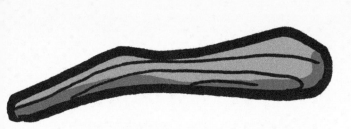

WOOD FOR THE WAND

FINE STEEL WOOL

**SANDPAPER
(60, 120, AND 220 GRIT)**

TUNG OIL

SELECT THE WOOD FOR YOUR WAND

Select a piece of wood that is slightly larger than the size of your desired wand. You can use a handsaw or dovetail saw to cut a dry, solid tree branch that you found on the ground to size. Or you can go to a craft or lumber store and buy a blank. A 12-inch blank of hardwood from the craft store is a great starting place.

BLANKS

Lumber stores and craft and hobby shops usually carry an assortment of wood that has already been cut into sections that range from 6–20 inches long. These pre-formed shapes are ideal for creating a straight, heavy wand, though you will have to sand it down to the right diameter. Some stores may do this for you.

ROTARY TOOL

A rotary tool is a handheld power tool that can be used for sanding, grinding, and polishing. A clamp called a collet holds different bits—or removable pieces that do a variety of things—in place at one end, making it very versatile!

To go with your rotary tool, I recommend the following bits:

★ ½-inch drum sander mandrel
★ 60-grit to 240-grit sanding band
★ ⅜-inch, 80-grit flapwheel sander
★ sanding discs
★ abrasive brushes
★ abrasive buffs
★ engraving bits

Use the rubber drum mandrel, which is one of the grippers that goes on your rotary tool, to put on a sanding band. The screw at the tip will expand the mandrel to make the drum fit tightly. If your wand is being made from a tree branch, start sanding at a low speed until you get the feel for it. Make light passes instead of trying to take all the bark off in one go.

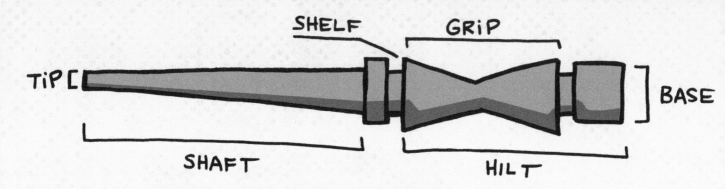

STEP 2: DESIGN YOUR WAND

Trace the shape of the wood on a large piece of paper (old grocery bags are perfect). Lay the wood on the paper and trace the outline.

★ Hilt

The hilt is the part of the wand you hold onto. It should be the widest part of the wood, and it should be designed to give you a comfortable grip.

★ Shaft

The shaft stretches from the hilt to the tip. The wand should taper a bit so that the tip is thinner than the base. The shaft can feature whatever designs your imagination can come up with.

★ Features

Draw lines on the traced sketch to indicate where you will carve and sand away wood to form your unique design features.

Create several designs using different features or spacing until you are satisfied with the look.

STEP 3: STRIP THE BARK AND SMOOTH THE SURFACE

If you are using a tree branch as the basis for your wand, you will need to strip the outer bark to expose the sapwood beneath.

Carve away any bumps unless you want to keep them as part of the wand's design.

Following the grain of the wood, sand off the layers of bark with 60-grit sandpaper until the sapwood is exposed.

As you get closer to the sapwood underneath, switch to higher grit to smooth the surface.

MAKEY'S TIPS!

FOR SOFTER WOODS, START WITH HIGHER-GRIT SANDPAPER!

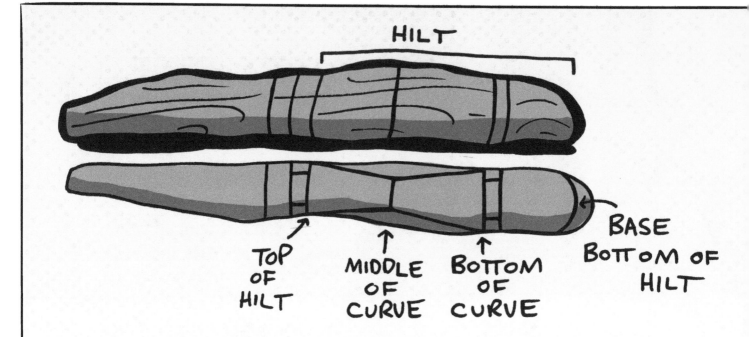

HILT

TOP OF HILT

MIDDLE OF CURVE

BOTTOM OF CURVE

BASE
BOTTOM OF HILT

STEP 4: SHAPE THE WAND

Update your sketch by tracing the shape of the exposed wood onto the same paper that you used for your original design. Add design features that you want to keep to the new tracing, making adjustments for width and shape. Using this new sketch as a guide, mark the wood where you'll be making cuts.

Now you can start carving the designs into the wood!

CARVING TIPS FOR THE HILT OF THE WAND

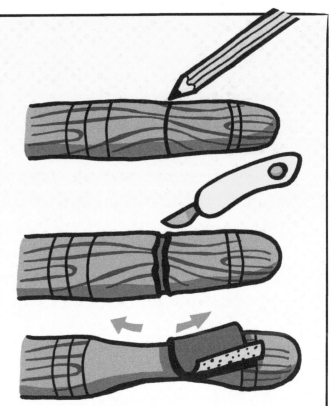

To add a **concave—or inward—curve** to the hilt, draw a line around the wand in the middle of the space marked for the grip. Following the line, cut a shallow groove all around the wand. Create a tapering, upward slope toward the either side of the grip by sanding on either side of the center groove. When both sides of the curve have been shaped, lightly sand the grip until you have a gradual, shallow curve.

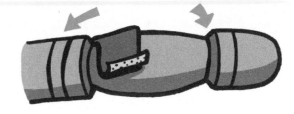

To make a **convex—or outward—curve**, carve a groove at either end of the grip and shape the slope upward toward the center.

To create a **ring** around the wand, mark both sides of the ring. Cut a trim line on the outside of the ring, following the marks using a small handsaw or your knife carving. Carve and sand down the outside wood ring.

MORE CARVING TIPS FOR FEATURES

To create a **square feature**, mark it on either side so that it's just as wide along the hilt as the wand is wide. Trim away the edges to the desired depth until you are left with a wide ring around the wand. Sand one edge of the wide ring to flatten it. Turn the wand over to place it on the flattened section, and flatten the opposite side. Then flatten the remaining two sides.

To create a **spherical feature**, create the wide square as above, and then carefully trim the edges and corners to form the rounded shape.

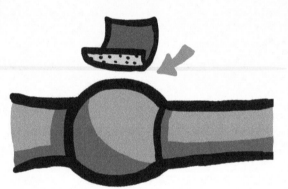

To make a **short gap**, cut two lines that go completely around the wand and then cut away the remaining wood in between them. This is easiest to do using the sanding bit on a rotary tool. If the gap is wide enough, peel away the remaining section with a ¼-inch sanding band or an engraving bit.

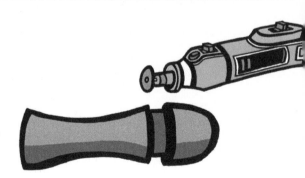

STEP 5: SMOOTH THE WOOD

Start with 60-grit sandpaper to remove the roughest patches along the entire wand. Make a second pass using 120-grit sandpaper to further remove uneven patches and to begin smoothing the wood. Continue to sand with 220-grit sandpaper until it's as smooth as you want it to be.

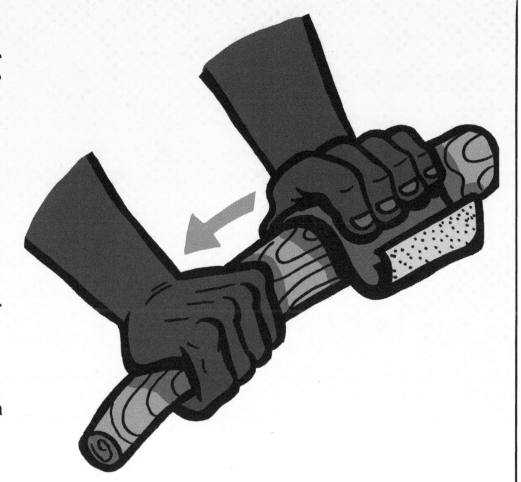

Sand *with* the grain to reduce the amount of fine scratches that will have to be removed later. The only time you should sand against the grain is when sanding the area around features—such as a ring—that have an edge where they meet the body of the wand.

All About Sandpaper

Sandpaper is made by coating a backing paper with glue and then putting down a layer of abrasive material. The size of the abrasive material is measured in "grit."

When you are deciding which sandpaper to use, remember: the lower the grit, the coarser the sandpaper. When you sand something, you remove all the scratches up to the size of the grit you're using. With rough wood, you might begin by sanding it with 80-grit sandpaper. You could then use 100-grit to sand out the scratches left by the 80-grit. If you keep repeating with higher and higher grits, the wood surface will become smooth. For most woods, 220-grit gives you a surface that looks and feels very smooth.

CAUTION!

Keep electronic devices such as smartphones away from the work area when you use steel wool.

STEP 6: FINISH THE WAND WITH OIL OR VARNISH

When the shape of the wand is complete and the wood has been smoothed, you can apply a coat of oil or varnish to protect the wood and finish it like a pro. Tung oil is great for wands.

Clean the wand's surface with a dry cloth to remove sawdust particles. Use a can of pressurized air or blow through a straw to remove sawdust from difficult areas.

APPLY AND DRY THE OIL

With a small brush or cloth, paint a thin coat of oil over the entire surface of the wand. Let the wand dry. To minimize contact with the surface, tie two strings around a horizontal board or pole and hang the wand between the two loops, or carefully lean it in an out-of-the-way corner. When the oil is dry, use fine steel wool to buff away the shiny topcoat.

TAKE IT FURTHER

You can enhance your wand even further by adding gems from costume jewelry, stones, beads, coins, and other small decorative items. You can simply glue them to your wand, but a better way to add decorations is to carve holes—or insets—for the items into the wand and then glue them into the insets.

CARVING AN INSET

Draw the outline of the item on the wood. Use a carving knife to carefully remove a shallow depth of wood from within the outline, starting in the center and moving outward to the drawn outline. Once that is done, glue the item into place. Use a cloth to wipe away any extra glue.

Essential Tools

For best results and to avoid injuries, always use the right tool for the job. Here are a few tools and related items that all great woodworkers have in their sheds.

★ Hammer or Mallet

You probably have a hammer, but a rubber mallet is better because it is less likely to dent the surface of the wood. Tip: If you use a regular metal hammer, put a "sacrificial" piece of wood in between to protect the good board.

★ Dovetail Saw/Coping Saw

These are much smaller than normal handsaws and are used when you need finer control to cut wood joints such as tenons or finger joints. They are also useful for cutting dowel rods.

★ Handsaw

Another common tool that you probably already have is a standard handsaw. It will be useful for breaking down larger boards and plywood.

★ Speed Square (or Triangle Square)

Woodworkers use this triangle-shaped tool to mark angles for cutting lumber. With it you can quickly determine if a corner is square (90 degrees). Squares come in a range of sizes; 6 inches is the most common.

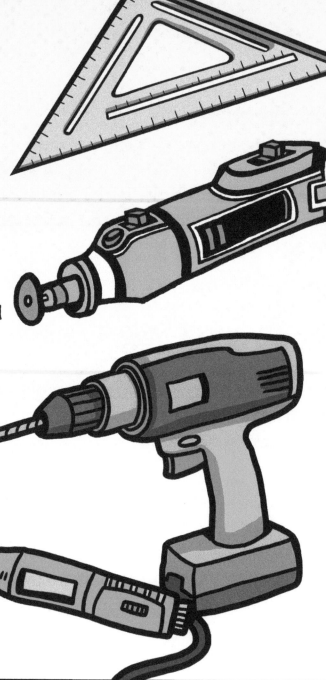

★ Rotary Tool & Bits

If the idea of carving and shaping wood appeals to you, then a rotary tool is a great purchase. It can also serve as a drill for small bit sizes up to 1/8 inch.

★ Electric Drill & Bits

A rechargeable electric drill (either a small pen-type model or a larger, pistol-grip type) is useful in hundreds of different situations. A standard bit assortment up to 3/4 inch will cover all your needs for the projects we'll be doing in this book.

★ Work Surface

Every woodworking Maker needs a surface to work on for clamping, drilling, sawing, gluing—you name it—and the kitchen table isn't it!

A piece of plywood across two sawhorses makes a solid workbench for basic projects. If you want to step up your woodworking game, a used table from a neighborhood garage sale or an inexpensive multipurpose workbench—such as the Wilmar W54025—that you can find online work great, too.

★ Toolbox

You might want to consider getting an inexpensive plastic toolbox. It's a great way to keep your tools together, organized, and in good shape.

MAKE A
SANDING BLOCK

•• ━━━━━━━━━━ ••

If you're going to be a woodworker, you're going to use sandpaper. A lot! ⚡
Holding it in your hand works really well for sanding curved objects, but when
you want to sand a board, a sanding block will save your tired hands. ⚡ This is
a simple project that teaches you skills you should master before trying more
complicated projects.

TOOLBOX

HANDSAW

SANDPAPER

LOW-TACK SPRAY ADHESIVE

SPEED SQUARE

PENCIL

MATERIALS

**PINE 2X4
(LONGER THAN A FOOT)**

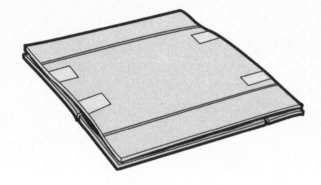

**SCRAP
CARDBOARD**

STAY SAFE!

When you use spray adhesive, do so outside or in an o pen garage or other well-ventilated area. Avoid breathing the fumes.

MAKE A SANDING BLO

STEP 1: CUT THE SANDPAPER

Measure and cut an 8 1/2-inch x 3 1/2-inch strip from a sheet of 100-grit sandpaper. You should be able to cut three strips of this size from a normal sheet of sandpaper.

STEP 2: MARK THE CUTTING LINE

Use your Speed Square to make a line 3 1/2 inches from the end of a pine 2x4. Butt the square up against the edge of the 2x4 so the square is parallel to the end of the board.

STEP 3: SECURE THE BOARD TO BE SAWED

If you have a workbench with a clamp, clamp the 2x4 to it so that the marked end hangs over the edge of the workbench. Otherwise, hold it firmly against the work surface with your hand.

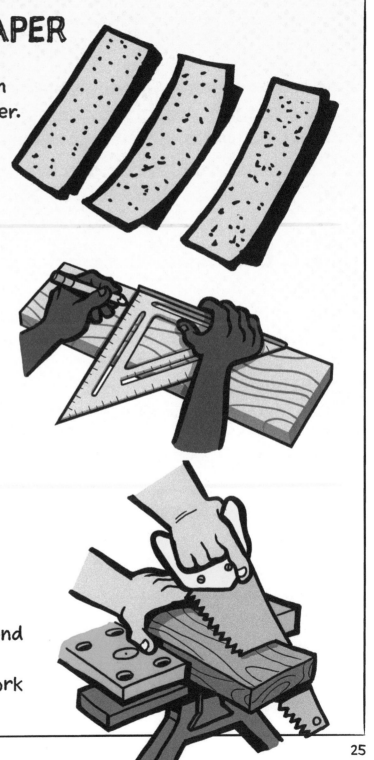

STEP 4: SAW THE BOARD

Using your handsaw, cut the end off of the 2x4 along the marked line. Make a few very light passes with the saw at first to create a groove in the wood. Once the saw is firmly on line, you can saw more aggressively.

STEP 5: SAND THE EDGES OF THE BLOCK

Take one of the strips of sandpaper, spray the back of it with low-tack adhesive, and attach it to a piece of scrap cardboard (an old pizza box lid works fine as long as it isn't greasy). Use this to sand the sharp edges of the block until they're smooth.

MAKEY'S TIPS!

LOW-TACK (STICKY, BUT NOT TOO STICKY) SPRAY ADHESIVE IS VERY USEFUL FOR HOLDING DOWN ANYTHING YOU DON'T WANT TO BE IN PLACE PERMANENTLY.

Easing an Edge

A good saw leaves a very sharp corner where it cuts. If you have sharp corners on a sanding block, they will quickly cut through the sandpaper as you use it. Removing these sharp edges is called "easing" the edge. Use a rolling motion with the edge while running it across the sandpaper. It gives you a gentle curve instead of a sharp corner.

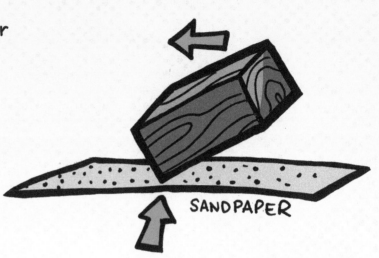

SANDPAPER

STEP 6: GLUE THE SANDPAPER TO THE BLOCK

Spray the back of the second piece of sandpaper with adhesive. Wrap the strip around the block so that the gap between the ends of the sandpaper is centered across the wide side of the 2x4 block.

STEP 7: PRESS THE SANDPAPER IN PLACE

Firmly push the sandpapered sides against the flat surface of your work surface. Your sanding block is now ready to use. When this piece of sandpaper becomes too worn to work, peel it off and replace it with another piece.

MAKEY'S TIPS!

FOR SMALL SANDING JOBS, YOU DON'T NEED TO SPRAY-MOUNT THE SANDPAPER—JUST WRAP THE SANDPAPER AROUND THE BLOCK AND HOLD IT IN PLACE WITH YOUR HAND.

TAKE IT FURTHER

Make a block for each of the common grits you will use—80, 100, 120, 150, 180, and 220. Label each of them with a permanent marker. After you've learned how to make a box (see page 56), you may want to make a storage crate for your sanding blocks.

Skillbuilder

If you want pro results in your woodworking projects, you've got to work like a pro. One really important thing that pros do is correctly mark and cut wood. It sounds pretty basic, right? It can be tricky. There are two types of lines to use to mark a line to cut: 1) a take line (TL) and 2) a leave line (LL). It doesn't matter which one you choose as long as you use the same type of line throughout a project. Remember to add a TL or LL next to the line so the person cutting knows what kind of line it is. Here's the difference between the two:

★ Take Line

When you mark a take line, it means that the blade of the saw should come down exactly on the mark. The goal is to perfectly cut away the marked line.

★ Leave Line

This means that you should cut just outside the line. When the section of the piece is cut off, the pencil line remains on the edge of the good piece.

All About Glue!

Without wood glue even the best woodworkers wouldn't be able to keep many of their pieces from coming apart. Yes, it's that essential. Other kinds of glue can be useful for bonding non-wood items to wood, but wood glue is key to bonding wood to wood, making joints stronger. Now that you respect the wood glue, here are some helpful tips for working with it.

★ The Glue Up

The final assembly of a project is called the "glue up." When applying wood glue to a joint, put a small line about the thickness of a BB down its length. Then use a slightly dampened finger to spread the glue across the wood. Apply glue to both of the pieces that are being joined together.

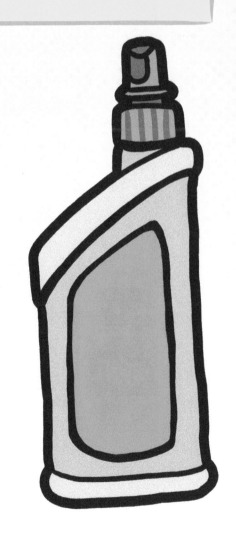

 Squeeze-Out

When the two gluey surfaces are pushed together, some of the glue will squirt out from between them. This is known as "squeeze-out." If you use just the right amount of glue on each side, you will get a tiny amount of squeeze-out. If you use too much, the glue will squirt out like a hot marshmallow slammed between two hammers!

There are two ways to remove squeeze-out:

★ Use a paper towel dampened with a little warm water to wipe away extra glue immediately after you've finished clamping the pieces together. You will want to make at least three passes to be sure you don't leave any residue. (Wood that has glue on it looks different under a finish than bare wood.)

Or

★ Don't touch it for 3-4 hours. At that point, the glue will have hardened and you can clean it off with a sharp chisel. If it is still kind of gooey when you start, stop and let it sit for another hour or two. Peace out!

REMEMBER!

With great power comes
great responsibility.

MAKE A
CHARGING STAND

A charging stand is a place where you can put your cell phone or tablet to recharge the battery. ⚡ This design is flexible and can be changed depending on your needs. ⚡ A typical tablet and phone can fit side by side on an 11-inch-wide board. ⚡ If you want a stand just for your phone, a 5-inch-wide board is plenty.

TOOLBOX

HANDSAW OR DOVETAIL SAW

POWER DRILL

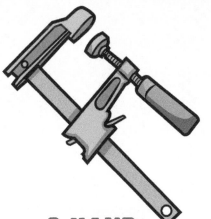

2 HAND CLAMPS

HAMMER OR MALLET

SPEED SQUARE OR T-SQUARE

PENCIL

$3/8$-INCH DRILL BIT

MATERIALS

WOOD GLUE

SANDPAPER (150, 180, AND 220 GRITS)

DIMENSIONED HARD-WOOD (SEE CUT LIST)

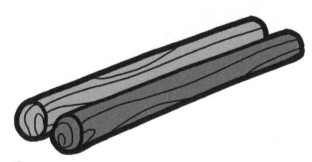

3/8-INCH WOODEN DOWEL STOCK (LIGHT OR DARK)

Cut List

A cut list tells you how to break down a board into the pieces you need.

PHONE + TABLET VERSION:

★ Piece A, Front: 11 inches x 11 inches x 3/4 inch

★ Piece B, Front Lip (the tablet and phone sit on it): 11 inches x 1 1/2 inches x 3/4 inch

★ Piece C, Back Stand: 11 inches x 5 inches x 3/4 inch

PHONE-ONLY VERSION:

★ Piece A, Front: 5 inches x 11 inches x 3/4 inch Or you can make it shorter: 5 inches x 7 inches x 3/4 inch

★ Piece B, Front Lip (the phone sits on it): 5 inches x 1 1/2 inches x 3/4 inch

★ Piece C, Back Stand: 5 inches x 5 inches x 3/4 inch

Dimensioned Lumber

Dimensioned lumber refers to wood that has been made square (all the edges are 90 degrees), planed to a specific thickness, and sanded smooth. (We use 1/2-inch and 3/4-inch lumber in this book.)

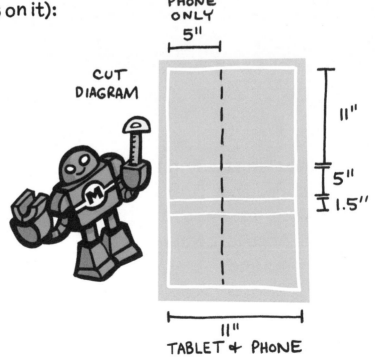

CUT DIAGRAM

PHONE ONLY 5"

11"

5"

1.5"

11"
TABLET + PHONE

STEP 1: MARK AND CUT THE FRONT (PIECE A)

If you have a 12-inch Speed Square, it's easy to make a straight line across the entire width of the board. If you're like me and you only have a 6-inch square, here's how to get around it:

⭐ Make a mark at 11 inches on each edge of the board

⭐ Use a straight edge (a yardstick or other long ruler) to connect the two marks with a pencil line

Remember: Always measure *twice* before cutting!

After marking the line, make a notation—either TL or LL—so you'll know where to start your cut (see page 30). Now cut along the marked line.

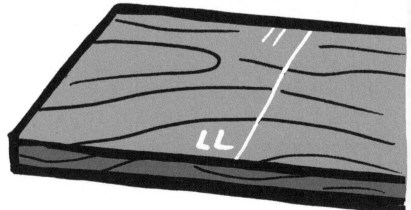

When you pick which side of the front (piece A) you want to be facing forward, write "front" on it, with an arrow pointing toward the top. This will immediately let you know which way the piece faces and which way is right-side-up.

SAWING A WIDE BOARD CAN BE TRICKY IF YOU TRY TO DO IT ALL AT ONCE. INSTEAD, START SAWING LIGHTLY AT ONE OF THE CORNERS TO MAKE A GROOVE BEFORE GRADUALLY TAKING A LONGER STROKE. HOLD THE SAW AT A 30- TO 45-DEGREE ANGLE TO THE SURFACE OF THE BOARD.

MAKEY'S TIPS!

USE A SMALLER BIT (1/16-INCH OR SO) TO MAKE A STARTER HOLE PERFECTLY CENTERED ON YOUR MARK. THEN SWITCH TO A 1/8-INCH BIT TO ENLARGE THE HOLE, THEN A 1/4-INCH BIT. NOW FINISH UP WITH THE LARGER 3/8-INCH BIT AND NOT HAVE TO WORRY ABOUT THE HOLE BEING OFF CENTER.

STEP 2: MARK AND CUT THE BACK (PIECE C)

Mark a line at 5 inches as you did in Step 1. Cut.

STEP 3: MARK AND CUT THE LIP (PIECE B)

Mark a line at 1 1/2 inches just as you did in Step 1. Cut.

STEP 4: MARK THE LOCATION ON PIECE A FOR THE BACK

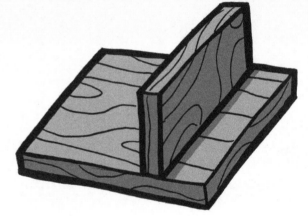

Make a mark on each side of the back of piece A, 2 3/4 inches from the bottom. Position the back (piece C) where it lines up with those marks, and then trace either side of it with your pencil.

After you've marked where the board sits on the back, make two marks in the center, about 2 inches from each edge. This is where you will put the dowels.

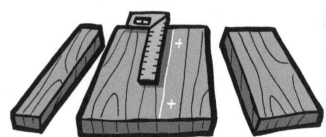

STEP 5: DRILL HOLES IN PIECE A

Now we're going to drill holes in piece A for the dowels. Start by placing a scrap backing board on to your work surface and then placing piece A on top of it. Then clamp piece A and the backing board firmly to the workbench. The backing board will not only keep you from drilling into your work surface, but it will also help keep the piece A wood from "tearing out" as the drill bit breaks through the other side.

STAY SAFE!

Always wear eye protection when using a drill and always clamp the piece you're drilling to a backing board—especially if you're drilling a large-diameter hole into a smaller piece of wood. A spinning drill bit can cause a piece of wood to twist out of your grip and become a whirling piece of destruction!

STEP 6: MARK AND DRILL THE BACK PIECE

Now that we know the location of the dowel holes, it's time to mark them onto the edge of piece C, the back stand. If you have a workbench with a vice, you can clamp piece C in the vice (use a couple of pieces of scrap wood to protect it from the jaws).

If you don't have a vice, grab a couple of bricks or cinder blocks and use them to support the wood. Place piece A over the edge of piece C, lining it up with the lines you drew earlier, and use a pencil to mark the center point of the two holes onto piece C.

Once marked, repeat the same sequence of drill bits on piece C that you used to drill out the holes in piece A—1/16-inch, 1/8-inch, 1/4-inch and finally 3/8-inch. Drill to a depth of about 3/4-inch to 1 inch, but this measurement isn't critical. We will cut the dowels to match the holes.

42

STEP 7: MARK AND CUT THE DOWELS

Now that we have all the holes drilled, it's time to cut the dowels that will hold the front and back pieces together. Start by placing the dowel rod into one of the holes in piece C. Use your pencil to mark where the edge of the hole meets the dowel.

Remove the dowel from the hole. Hold it up to the edge of piece A, and add a second mark to add the thickness of the board to the dowel.

Use your dovetail saw or coping saw to cut the dowel at the second pencil mark. It's better to cut a little too long than a little too short. The dowel can be sanded flush later.

Repeat for the second hole. As you measure the dowels, mark both the holes and the dowels so you know which dowel goes where once you make the cuts. You should now be able to test-fit the back to the front with the dowels in place and see the beginning of a stand!

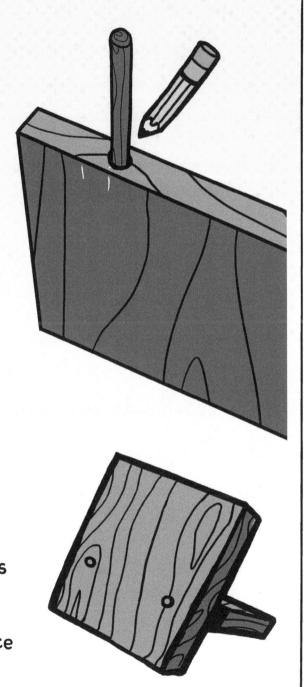

STEP 8: MARK AND DRILL THE FRONT LIP AND DOWEL

We're only going to use a single dowel to help support the front lip. Since the phone or other device doesn't weigh much, the combination of wood glue and a single dowel will be plenty.

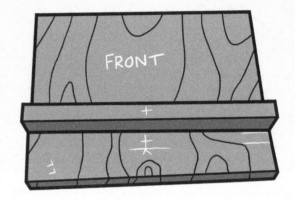

As you did with Step 4, make a couple of marks on the front of piece A, about 1 1/4 inches from the bottom. Trace the outline of the front lip. Mark a spot roughly centered between the lines and centered on piece A.

As you did in Step 7, mark the spot for the dowel on the edge of piece B. Make the hole for the dowel by working your way up through the drill bit sizes. Mark the depth of the dowel, add the width of the piece A board, and cut the dowel to length.

When it's time to test-fit, use a pair of clamps to hold the lip straight. It may spin around the dowel if you don't.

STEP 9: DRILL THE CABLE PASSAGES

Now that you have the entire piece together—yay!—set your phone or other device on the lip. Mark the location of each device's charging port on the lip.

Use your Speed Square to line up the marks on the lip (piece B) with the bottom edge of the front (piece A). Mark a location roughly centered for the hole.

Clamp the pieces to your backing board and drill out the holes for the cable pass-through. (A 3/8-inch hole should fit most cables.)

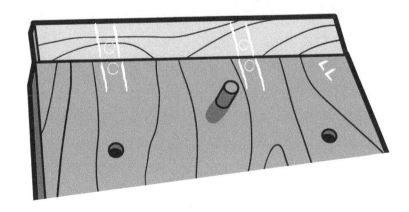

STEP 10: GLUE UP

Start by gluing the dowels into pieces B and C. Put a little bit on the bottom and sides of each dowel, and then slide it into its hole with a twisting motion so that the glue spreads evenly around the sides. Tap the dowels into the holes with a mallet. Let the glue dry for a couple of hours. (Go play with the dog!)

Next, use your finger to spread an even layer of glue along the edge of piece C, including the parts of the dowel that are sticking out. Slide the piece into place, inserting the dowels into the holes in piece A and tapping with a mallet, and use your clamps to firmly press pieces A and C together. If you have painter's tape, you may want to put strips of it down along the joint to protect against "squeeze-out" (see page 32).

Repeat this process for the lip—gluing, tapping, and clamping pieces A and B together—and let the glue dry overnight.

STEP 11: SAND THE STAND

If you started with dimensioned lumber, you already have a smooth base to start from. At a minimum, however, you will want to ease all the sharp edges (see page 27) and remove all the pencil marks. Use your sanding block, and work through grits 150, 180, and 220. If you started with raw lumber, you'll need to begin with 80-grit and work up from there.

STEP 12: FINISH WITH OIL

I recommend Danish oil for your first finishing job. Pour some on to a soft, lint-free cloth (an old t-shirt is perfect) and wipe it on generously. Let it sit for 5–10 minutes, then wipe it off with a clean cloth. Allow it to dry in a garage or other area where the smell won't annoy anyone.

STAY SAFE!

When you use a cloth to apply an oil-based finish, the oily cloth becomes a potential fire hazard. As it dries, heat is generated by the oil. To safely dispose of oily cloths, spread them out flat on the driveway or hang them completely spread out over a metal railing or fence. Once they have dried completely, they'll be stiff and safe to throw away.

Finishing Your Wood

When it's time to apply a finish to your projects, you have so many different options! Do you want to keep the natural color of the wood, or do you want to stain it a *different* color? Staining can be subtle, or it can completely change the look of the wood—especially if you start mixing your own colors!

Note: Be sure to read and follow the instructions that come with the particular product you buy.

★ Applicators

Most stains and finishes can be wiped on with a soft cotton rag. You can use an old t-shirt, or you can get a giant box of rags from the hardware store. You can also use a finishing brush or sponge, though these require a certain amount of care to clean after use.

★ Gel Stains

This is my favorite type of stain. They wipe on easily and don't run, and if you put on too much, you can just wipe the excess off.

★ Aniline Dyes

If you want brilliant colors that aren't found in nature (vivid blues, oranges, purples, and so on), aniline dyes are the ones for you. It takes just a tiny bit to turn a neutral finish into a brightly colored one, and the beauty of the wood still shows through.

★ Danish Oil

Danish oil is my default recommendation to someone who wants a fast, easy, hard-to-mess-up finish. Most Danish oils are a blend of oils, thinner, pigment, and drying agents. Apply it by wiping it on thickly, letting it sit for 10 minutes, and then wiping it off with a clean section of the rag. After that, it needs to cure for 48–72 hours. Danish oil has a very strong odor until it has finished curing.

★ BLO + Turpentine

If you get hooked on woodworking, you may find it cheaper to mix your own version of Danish oil. Mix a 50/50 proportion of boiled linseed oil (you don't have to boil it, they do that at the factory) with turpentine.

★ Water-Based Polyurethane

Polyurethane used to be tricky to apply, but that changed when they came out with water-based poly. Apply at least 3–4 coats to get a good, glossy finish.

Joinery: The Art of Holding Wood Together

The technique for getting pieces of wood to stick together is called "joinery." The most obvious way of joining pieces of wood together may seem to be with nails or screws, but most woodworkers take pride in never using anything but glue and wood. (Okay, there are some exceptions. We use screws to attach hinges.) Let's look at half a dozen of the most popular types of wood joints.

★ Dowel Joint

Tools needed:

★ drill with a bit the same size as the dowel

★ dovetail or coping saw

★ hammer or mallet

This is the only type of joint we'll make in this book. It is one of the

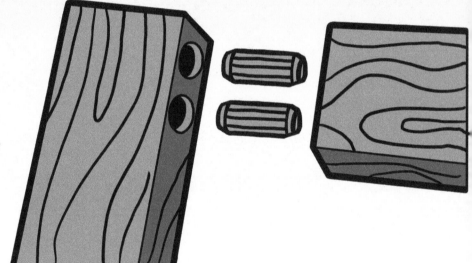

fastest and easiest methods of joinery, but it also happens to be one of the strongest. A dowel joint uses a cylindrical piece of wood cut from a longer stick. A hole is drilled through the pieces being joined, glue is applied, and the dowel is tapped down into the hole.

⭐ Butt Joint

Tools needed:

⭐ clamps

Okay, cut the butt jokes! This is the simplest way to connect two boards. Put some glue on each piece, butt them together, and clamp. The strength of a butt joint is largely dependent on how much surface area is glued.

★ Miter Joint

Tools needed:

- ★ corner clamps
- ★ miter box
- ★ dovetail saw

To join two boards together at a 90-degree angle, you can make a very strong joint by cutting each of them at a 45-degree angle and gluing them together. The best way to make the 45-degree cut is using a Speed Square or miter box. You need a special kind of clamp called a "corner clamp" to hold it together during glue up.

★ Lap Joint

Tools needed:

★ clamps

★ Speed Square

★ dovetail saw

A lap joint removes half the thickness of a board on each piece, then overlaps the pieces. It's useful when joining together wide pieces, like a picture frame.

★ Finger Joint

Tools needed:

- ★ clamps
- ★ Speed Square
- ★ dovetail saw
- ★ chisels

This joint uses interlocking "fingers" to join two pieces of wood together. This produces an attractive, very strong joint that can be manipulated and spaced any way you like.

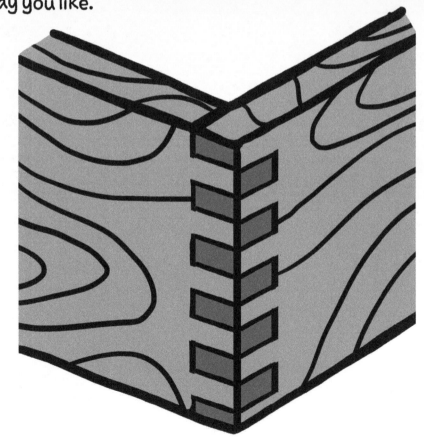

★ Mortise & Tenon Joint

Tools needed:

★ clamps

★ Speed Square

★ dovetail saw

★ chisels

This joint is very useful for shelving. Cut the tenon (the part that sticks out) in one board with a dovetail saw and the mortise (a rectangular hole for the tenon) in the other board with chisels. Accurate marking is important!

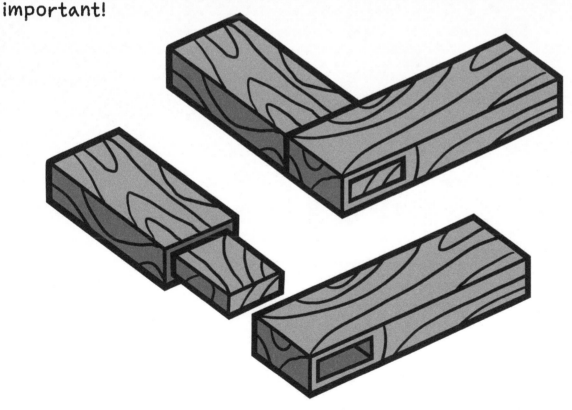

MAKE A SECRET-
COMPARTMENT BOX

•••—————————•••

And now for the finale! ⚡ We're going to make a wooden box, but not just any box. ⚡ This is going to be a box with a secret compartment! ⚡ Once you've mastered this project, you'll be able to design your own boxes—a jewelry box, a storage box for your sanding blocks, or maybe even a toolbox.

TOOLBOX

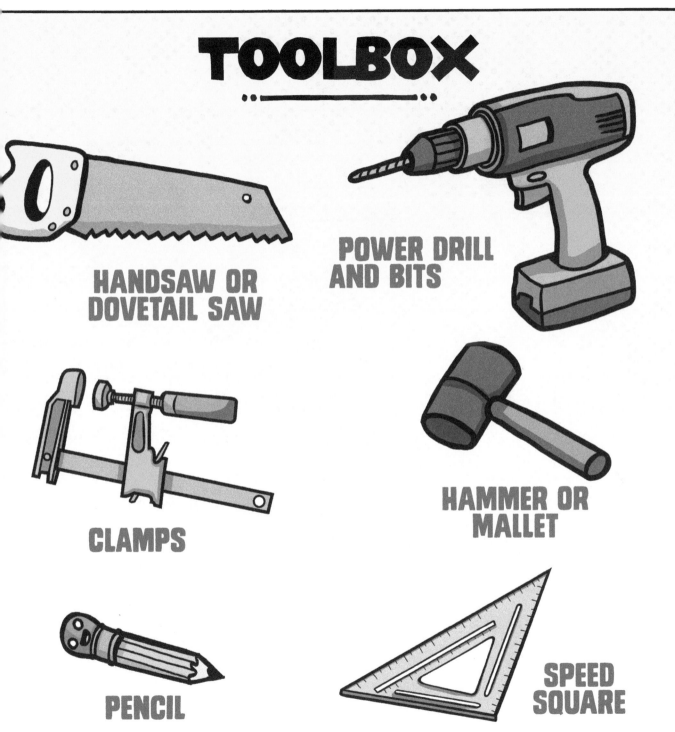

HANDSAW OR DOVETAIL SAW

POWER DRILL AND BITS

CLAMPS

HAMMER OR MALLET

PENCIL

SPEED SQUARE

MATERIALS

WOOD GLUE

SANDPAPER (150, 180, AND 220 GRITS)

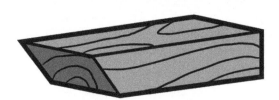

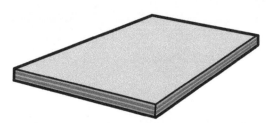

3/4-INCH THICK HARDWOOD

1/4-INCH THICK PLYWOOD (OR HARDWOOD IF YOU CAN FIND IT)

Cut List

⭐ Piece A, Real Bottom: 8 inches x 5 1/2 inches x 1/4 inch, plywood or hardwood

⭐ Piece B, False Bottom: 1/4 inch, plywood or hardwood; cut to measure later

⭐ Pieces C and D, North and South Sides: 8 inches x 5 1/2 inches x 3/4 inch, hardwood

⭐ Pieces E and F, East and West Sides: roughly 4 inches x 5 1/2 inches x 3/4 inch, hardwood; cut to measure later

⭐ Piece G, Lid: 8 inches x 5 1/2 inches x 3/4 inch, hardwood

⭐ Piece H, Lip for Lid: identical to piece B; cut to measure later

Skill Builder

My cut list is based on the 5 1/2-inch-wide and 1/4-inch-thick plywood that I found at the store. As long as you keep the wood thicknesses the same, you can vary the dimensions to suit yourself. If you're going to change them, however, write out your own cut list like the one above so that you don't accidentally mix up a number while you're cutting. With the charging stand, you could cut all of the wood ahead of time. Here, by contrast, you're going to use the first piece you cut to help mark the length of the second piece, and so on. Each step will tell you how to cut the next.

STEP 1: **START FROM THE BOTTOM**

A proper box is *square*. That doesn't mean it has to be literally shaped like a square. In woodworking when we talk about being square, we mean that all four corners of the box are exactly 90-degree angles. If your sides are "out of square," the top and bottom won't fit properly.

If you have a set of corner clamps, you'll be able to easily get a square layout. Because most people don't have corner clamps lying around, we're going to square the project from the bottom up.

Mark a square line across the 1/4-inch plywood. (I'm making mine 8 inches.) Make this a leave line, so mark it with an LL. This will make it easy to see what needs to be sanded away if the piece isn't square.

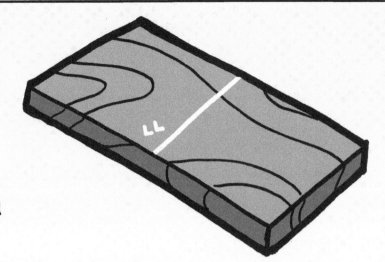

Piece A is the most critical cut you'll make for the project. After you cut it with your saw, use a straight edge to check all four corners to be sure they are square. If they aren't, use a sanding block with 100- or 120-grit sandpaper to bring them square. (See "Testing Square" on page 68.)

This is the piece that we'll use to keep everything square for the rest of the build. Choose one side and mark it to be the face-up side. Label the edges North, South, East, and West.

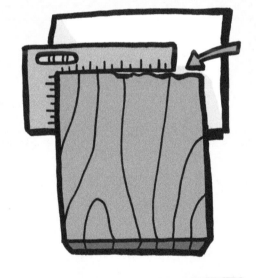

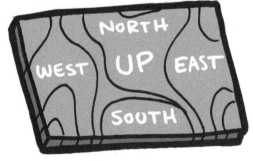

STEP 2: MARK AND CUT THE TOP

The lid (piece G) will be cut from thicker wood, but the other dimensions should be exactly the same as the bottom (piece A). Use piece A to mark the wood for the lid, and cut the lid out using your saw.

Put piece G aside for now. (Don't worry. We'll come back to it!)

STEP 3: CUT THE FIRST TWO SIDES

Use the north edge of piece A to measure the first side piece. Cut. Pick the side you want to face out and write "C – North – Out" on that side. This is now piece C.

Use piece C to mark the length to cut for the opposite board side. Cut. Pick the side you want to face out and write "D – South – Out" on that side. This is now piece D.

These two pieces should be identical in length. If they aren't, use your sanding block to make them the same. They should also fit perfectly along the edges of piece A.

STEP 4: CUT THE OTHER TWO SIDES

Using the east edge of piece A, mark the length for that side. Don't cut it, though. Hold pieces C and D together, and subtract their thickness from the line you just marked, as shown. Cut to the shorter mark.

Pick the side you want to face out and write "E — East — Out" on that side. This is now piece E. Use piece E to mark the length to cut for the opposite board side, just like we did in the last step. Cut.

Pick the side you want to face out and write "F — West — Out" on that side. This is now piece F. Piece E and piece F should be identical in length. If they aren't, use your sanding block until they are.

STEP 5: CHECK THE FIT

You should now be able to place each side on the bottom piece edge that matches it. The sides should be in square alignment at the corners.

STEP 6: MARK AND DRILL THE SIDES

If you made the charging stand project earlier in the book, you learned how to mark the boards and drill for dowels. We're going to repeat that process here. If you didn't make the charging stand, see pages 42–43.

Use piece E to mark where the dowels will go on pieces C and D.

After you've marked the board thickness, use your Speed Square to make a line going down the middle and mark targets for your drill bit to start the dowel holes.

Remember, start drilling with a small bit. Then gradually work your way up to the full 3/8-inch bit.

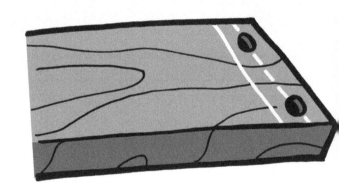

Use the holes in pieces C and D to mark the edges of pieces E and F for drilling.

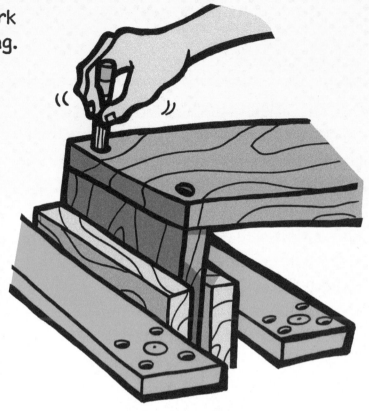

Use your series of drill bits to progressively drill the 3/8-inch dowel holes into the edges.

STEP 7: CUT THE DOWELS

Put the sides together again resting on the bottom (piece A). Starting with the south side (piece D), insert a dowel into one of the holes, pushing it all the way through into the edge of the piece. When the dowel is completely in, mark a pencil line around it flush with the surface. Cut the dowel at the mark. Repeat this for the other three holes. Now switch to north (piece C) and repeat the process for the 4 dowel holes on it.

When Dowels Go Bad

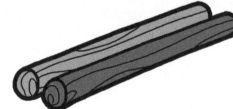

If you have trouble getting the dowels to line up, don't panic. Instead of pre-drilling for the dowel holes, glue up the box using butt joints (see page 51). After the glue has dried, clamp the box to your work surface, mark where the dowels go onto the front and back of the piece, and drill the dowel holes (working up from a small bit). After you've drilled the holes, apply glue to the sides of the dowels and then insert and tap them into place.

STEP 8: GLUE UP THE SIDES

Put a little glue into the dowel holes in pieces E and F, and then smear glue along the edges of the dowel. Slide them into place and tap with a mallet. When you're finished with all 8 dowels, clamp the opposite sides together for the glue to dry. Use the bottom (piece A) to make sure the sides are square. If you have enough clamps, you can even clamp the sides in place to the bottom. Let it dry overnight. Get some rest.

Come back tomorrow.

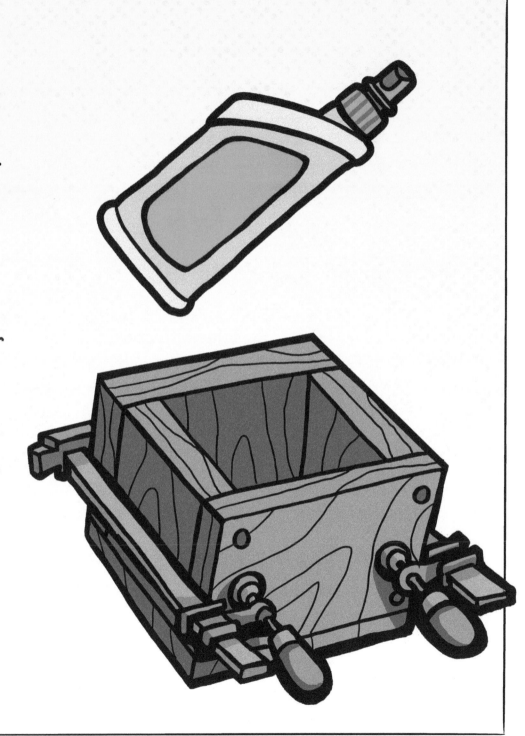

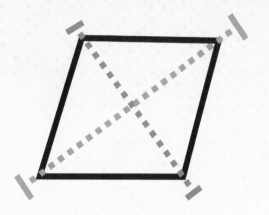 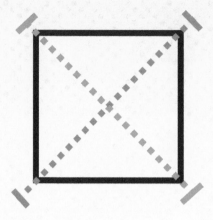

Testing Square

Place your square against one corner of the box. If there is a gap between the edge of the square and the side of the box, they are out of square. You can gently tap the side to close up the gap. Once you have one corner perfectly square, the others should also be fine (but check them anyway!).

You can also test for square by measuring the distance between opposite corners (the diagonals). If the measurement is equal, the corners are square. If not, you'll have to gently move the piece into alignment by moving one of the corners and measuring again until the numbers match.

Note: If you don't have a tape measure handy, you can use a piece of string to check the diagonals. The exact measurement isn't important. You just care that they are equal.

MAKE A SECRET-COMPARTMENT B

STEP 9: MARK AND CUT THE LIP AND FALSE BOTTOM

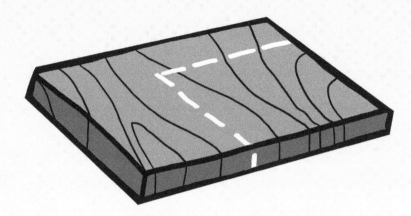

Now we're going to use the glued-up sides to mark a couple more pieces of 1/4-inch wood.

Caution: Handle the sides very gently until you get the bottom glued on.

Place the sides on top of the plywood, with one of the corners of the plywood just inside the walls of the box. The illustration shows how you want to wedge the corner of the plywood into the corner of the box. This will save you from having to cut out all four sides.

Use a pencil to trace the inside dimensions of the box onto the plywood.

Cut this piece with your saw. The resulting piece should just barely slide into the inside of the box with no visible gaps. If it's a little big, touch it up with sandpaper.

This is piece B, your false bottom!

Trace the outline of this piece onto the plywood, and then cut out a second identical piece. This is piece H, the lip for the lid.

STEP 10: GLUE THE LIP TO THE TOP

Remember piece G that we cut way back in Step 2 and set aside? It's time to bring it out again!

Put the side of piece G that you want to show as the lid face down on your workbench. Place the sides of the box on top of it so that they line up with the edges.

Take the lip you just made, piece H, and smear some wood glue on one side. Don't put any glue within 3/4 inch of the edge. We don't want to accidentally glue the piece to the sides of the box.

With the glue side facing down, lower piece H down in the center of the box until it is flat against the piece at the bottom. Place something heavy on the center of it (I used a small brick), and lift the sides away. When it dries, you should have a lid that perfectly nestles into the top of the box. Set it aside to dry.

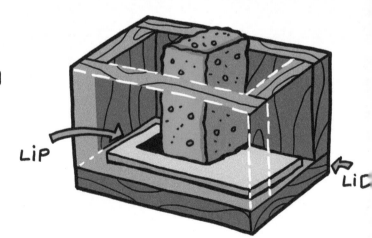

LIP

LID

STEP 11: DRILL KEY HOLE AND GLUE BOTTOM

Let's go back to piece A, the real bottom of the box. With your ¼-inch drill bit, drill a hole through the bottom a few inches in from one of the edges.

This keyhole is the "trick" behind the secret compartment. Insert a pencil or some other narrow object into the keyhole and lift up the false bottom to expose the secret compartment beneath.

Next, run a bead of glue around the bottom edges of the sides and smear it across the entire set of edges with your finger.

Carefully place the bottom onto the glued edges and clamp it on. Let it dry.

SECRET COMPARTMENTS ARE HANDY!

I KEEP AN EXTRA CAN OF OIL IN MINE!

STEP 12: CUT AND GLUE STANDOFFS

The final step before sanding and finishing is to add the standoffs for the false bottom. It's up to you to decide how tall you want to make the secret compartment, but ideally it should be no more than 25 percent of the height of the box. If it's any higher, then it will be obvious from just looking at the inside of the box that there's something "funny" about the bottom. For a box that is 4–5 inches tall, cut the standoffs to 1 inch.

You've probably got enough scrap offcuts to use to make the standoffs. They don't have to be the same in all dimensions. As long as they're all 1 inch tall, they'll work. Use your saw to cut these pieces.

Once your standoffs are cut, glue them into the bottom corners of the box.

STEP 13: SAND AND FINISH

If you made the charging stand, the finishing process is identical. Sand the surface with 150-, 180-, and 220-grit sandpaper. Apply oil, let it sit for a few minutes, and then wipe off the excess. Let the box cure for several days in a garage or someplace else where the smell won't bother anyone.

YOU CAN BUILD UP A DEEPER FINISH BY APPLYING MULTIPLE LAYERS A FEW DAYS APART. THE KEY IS TO LET EACH LAYER DRY COMPLETELY BEFORE APPLYING THE NEXT ONE.

MAKEY'S TIPS!

TAKE IT FURTHER

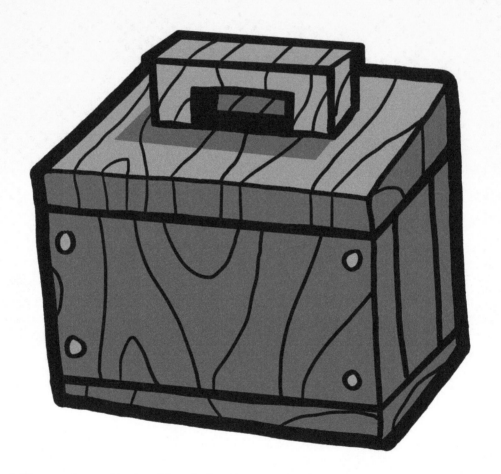

Consider adding a handle to the lid of your box. Use your carving knife or rotary tool to shape a piece of wood into a knob. Leave one side perfectly flat. Glue it to the lid. For an even stronger joint, drill a hole through the lid into the knob and insert a dowel.

Common Woodworking Woods

 Pine

Pine is the most readily available wood. It's cheap, and it looks good when properly stained and finished. The down side of pine is that it's very soft, so it dents and scratches easily. If you like things to look weathered and well-used, this could be an up side!

★ **Oak**

Red oak is the most common (and usually the least expensive) of the hardwoods used in woodworking. Its close cousin, white oak, is a bargain as well. It's a very durable wood that looks good whether stained or finished naturally.

 Cherry

Cherry wood is easy to work, and it is gorgeous looking whether finished naturally or with a subtle stain.

★ **Walnut**

With nothing but a clear stain, walnut wood gives you a deep, vibrant surface. It is, however, more brittle than cherry.

★ **Plywood**

Plywood can also be useful for a woodworker. For instance, 1/4-inch plywood makes a good box bottom. Look for furniture-grade plywood that is already sanded. The cheapest is birch plywood, a very light wood that stains well. (Construction-quality plywood, or plywood made from pressed wood, isn't suitable.)

Other woods include mahogany, maple, teak, cedar, and even exotics like zebrawood or cocobolo. Some people are allergic to the dust of exotic woods. It is good practice to always wear a dust mask when sanding any wood.

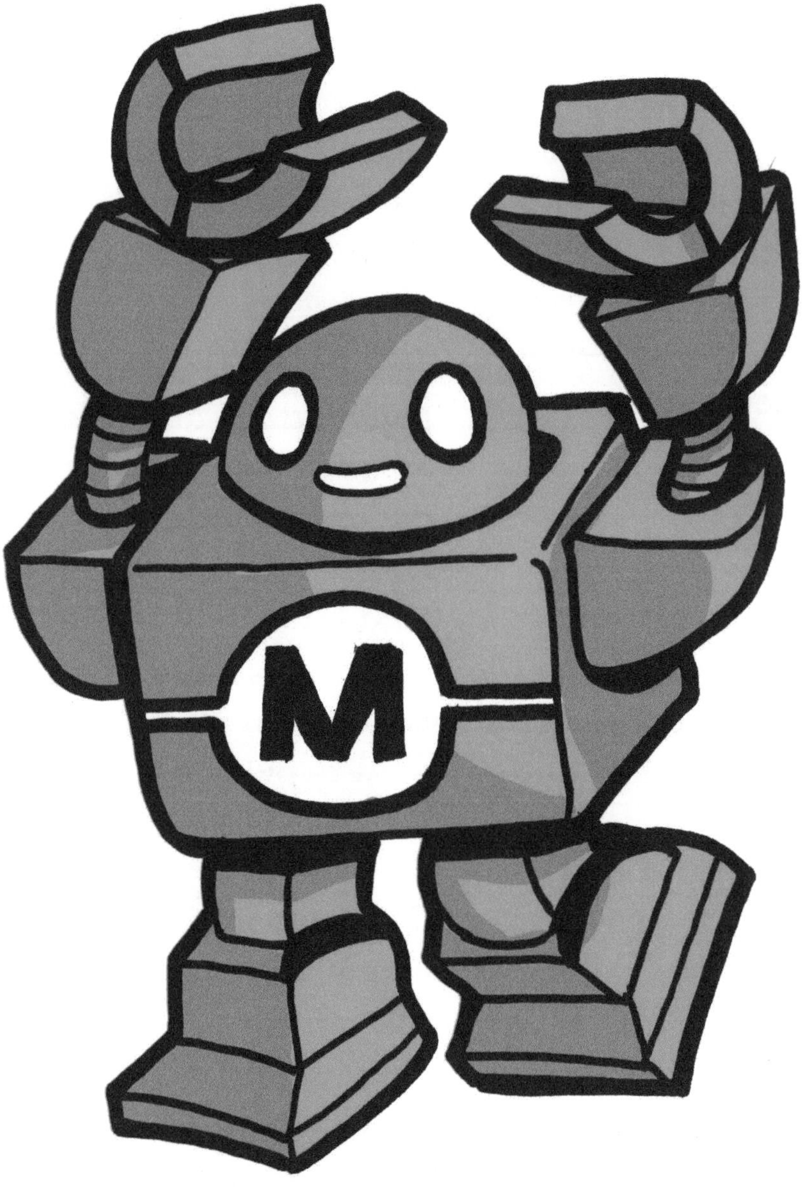
M

WHAT'S NEXT?

ongratulations! If you finished the projects in this book, you've taken
everal giant steps toward becoming a serious woodworker. What's even
more important is that you're now officially a Maker like me! If you'd like to
hare your projects with other Makers, visit our website and tell us about it
t makezine.com/contribute.

Voodworking is a popular hobby. You can find more information about it by
atching videos online and by visiting your local library or bookstore. Look for
Maker Faire in your area to find out what other enthusiastic Makers are
oing, and consider joining a makerspace in your community to learn from
xperienced Makers who are happy to share their knowledge with you!

Maker Faire ®

Maker Faire is the Greatest Show (and Tell) on Earth. It's a family-friendly festival of invention, creativity, and resourcefulness and a celebration of the Maker movement.

Maker Faires are held all over the world and are gatherings of people who love to share what they are doing. **Visit a Maker Faire in your area.** You'll meet other Makers like you and experience some of the amazing things they have created. You may even want to show off your own projects at a Maker Faire someday.

Find us online at MakerFaire.com to learn more about making and to find a Faire near you!

ALSO AVAILABLE

CPSIA information can be obtained
at www.ICGtesting.com
Printed in the USA
JSHW012032081220
10078JS00016B/87